6 31
2

library 2000

GREAT INVENTIONS
DISCOVERIES AND INVENTIONS
FAMOUS ARTISTS AND COMPOSERS
SHIPS
WEAPONS OF THE PAST
MODERN WEAPONS
EARLY AIRCRAFT
MODERN AIRCRAFT

Published in Great Britain by Frederick Warne (Publishers) Ltd, London

First published in Great Britain in 1980
Reprinted 1984

ISBN 0 7232 2426 9

Phototypeset by Tradespools Ltd., Frome, Somerset

Printed in Spain by Sirven S.A.E., Barcelona
Depósito Legal B. 13.050-84

library 2000

Written and illustrated by Vincent Segrelles

Translated by Arthur Butterfield

Modern Weapons

FREDERICK WARNE

CONTENTS

INTRODUCTION

Man today has reached the peak of his destructive power. Between the first weapons of war and those in use today there is a measureless gap. Human progress has served not only to benefit mankind but also to face it with destruction. In the last two wars deadly weapons have achieved the capability of massive extermination. Man is faced with the stark choice of living or of being wiped out. There is no doubt that the balance of terror has helped to delay the start of World War III. While managing to evade total annihilation, men have nevertheless embarked on a series of wars using conventional weapons. Can war itself not be wiped out? Has man always got to fight his fellow man? Why? What is man's destiny? Is man faced with a permanent peace based on the balance of terror, or will he be plunged into a new war?

This history of the tools of modern warfare covers the development of automatic weapons over a hundred years ago, World Wars I and II, and the postwar period.

THE METAL CARTRIDGE

The metal cartridge incorporating its own primer was thought up more than a hundred years ago by the American gunsmiths, Horace Smith and Daniel B. Wesson. At about the same time the metal cartridge was also developing in Europe, but in a slightly different way. This used a needle or firing-pin, instead of having the primer in the middle of the base of the cartridge. The new type of cartridge was used first for revolvers, and in this way an old problem was solved. It should be remembered that the revolver was a percussion-cap weapon and was a muzzle-loader. That is to say, the breech or chamber was covered up at the rear and had only a tiny opening through which to insert the fulminant. Once the fulminant (explosive salt), the powder, and the bullet were all housed in one casing, the way was open for the development of automatic weapons.

In North America the manufacture of revolvers was monopolized by Colt. Colt's patent lapsed in 1857 and at that date Smith and Wesson had ready their new model with a chamber bored to take the latest cartridge. In Europe, Lafaucheaux had patented his needle cartridge. The Smith and Wesson system was extremely successful in the Far West and was adopted as the official weapon by both Wells Fargo and American Express. But Colt in the meantime had not been idle, and in 1873 he produced a new metal cartridge case which was so successful that from that moment the name of Colt was identified with the revolver. This was strange because both Remmington and Smith and Wesson produced revolvers that were better than the Colt.

In the picture below you can see a Smith and Wesson revolver opened to remove the empty cartridges. In order to make it clearer, there are just two cartridges in the chamber which normally takes six.

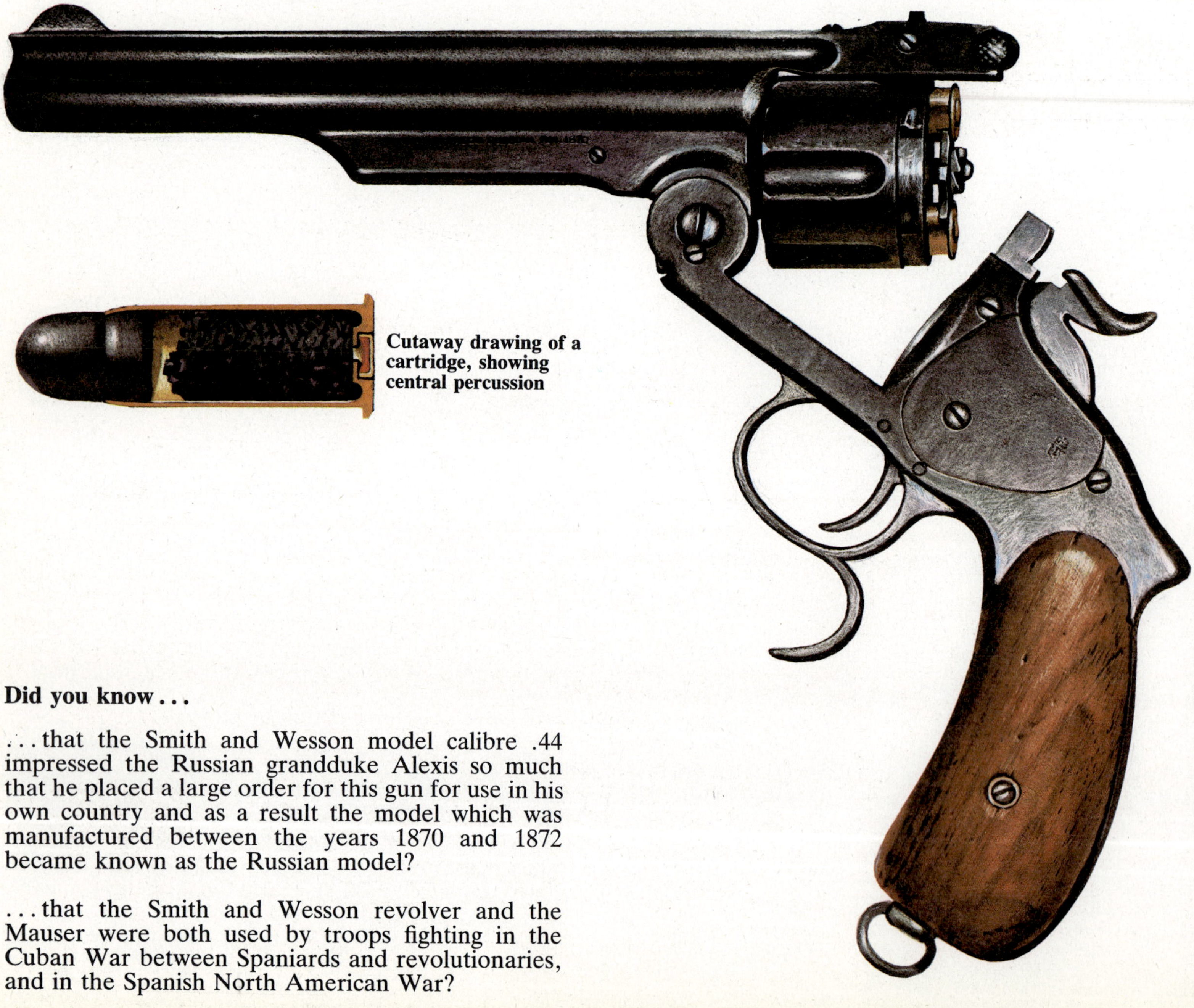

Cutaway drawing of a cartridge, showing central percussion

Did you know . . .

. . . that the Smith and Wesson model calibre .44 impressed the Russian grandduke Alexis so much that he placed a large order for this gun for use in his own country and as a result the model which was manufactured between the years 1870 and 1872 became known as the Russian model?

. . . that the Smith and Wesson revolver and the Mauser were both used by troops fighting in the Cuban War between Spaniards and revolutionaries, and in the Spanish North American War?

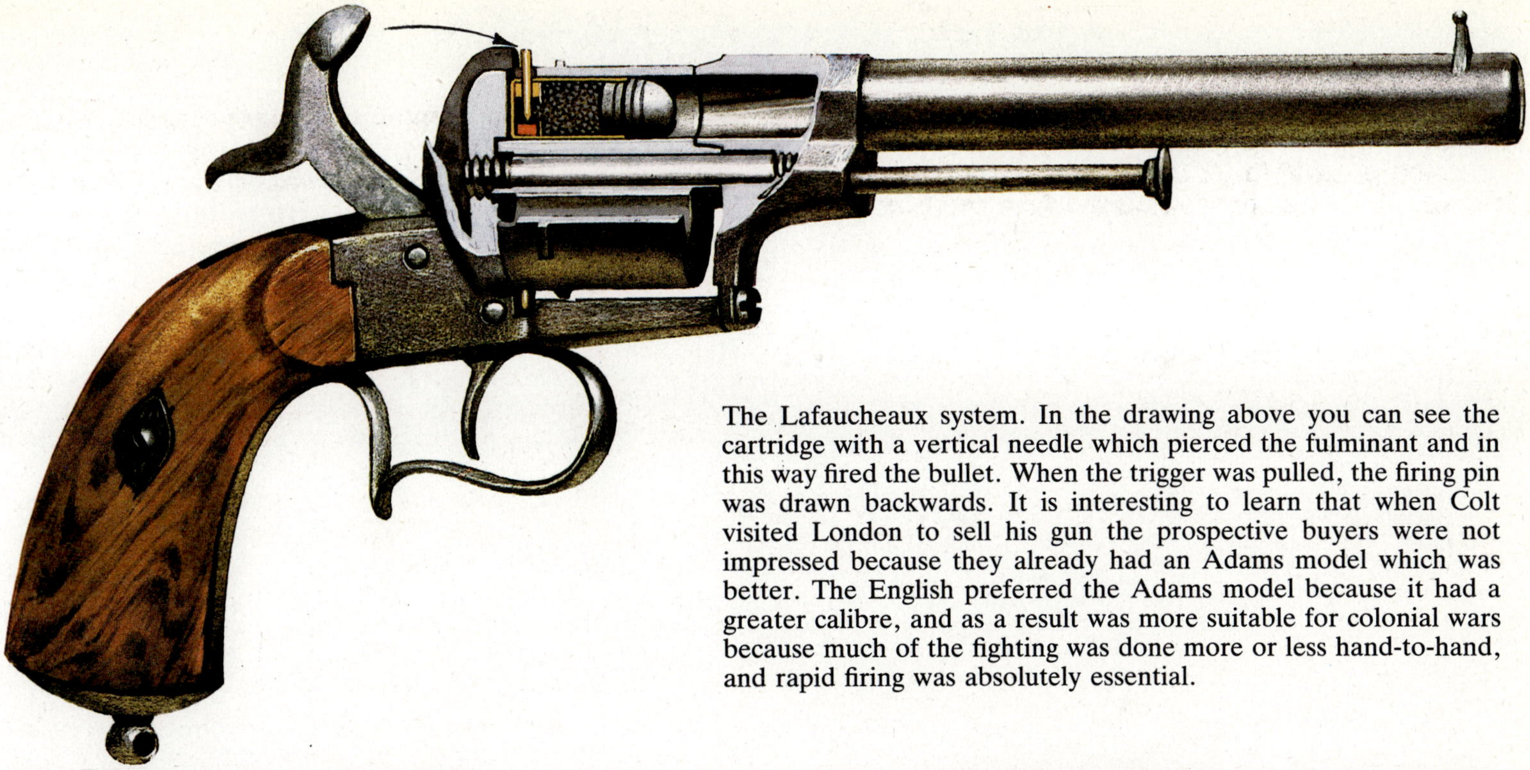

The Lafaucheaux system. In the drawing above you can see the cartridge with a vertical needle which pierced the fulminant and in this way fired the bullet. When the trigger was pulled, the firing pin was drawn backwards. It is interesting to learn that when Colt visited London to sell his gun the prospective buyers were not impressed because they already had an Adams model which was better. The English preferred the Adams model because it had a greater calibre, and as a result was more suitable for colonial wars because much of the fighting was done more or less hand-to-hand, and rapid firing was absolutely essential.

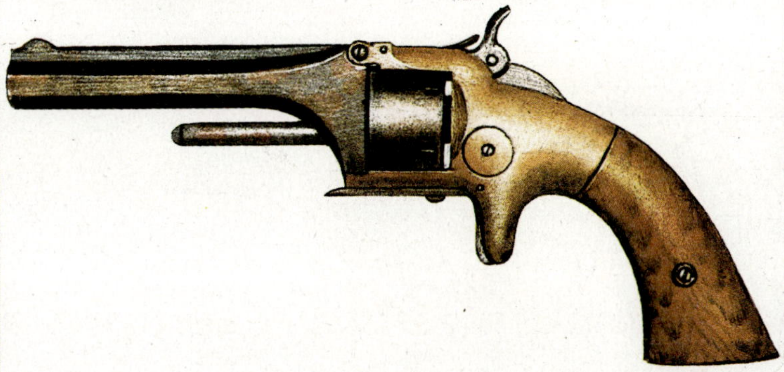

The first Smith and Wesson revolver is shown here: it was a .22. It cost $12, and 100 cartridges cost 25c. More than 126,000 of these weapons were made in ten years. Another model, the .32, was also extremely successful and was much used during the American Civil War.

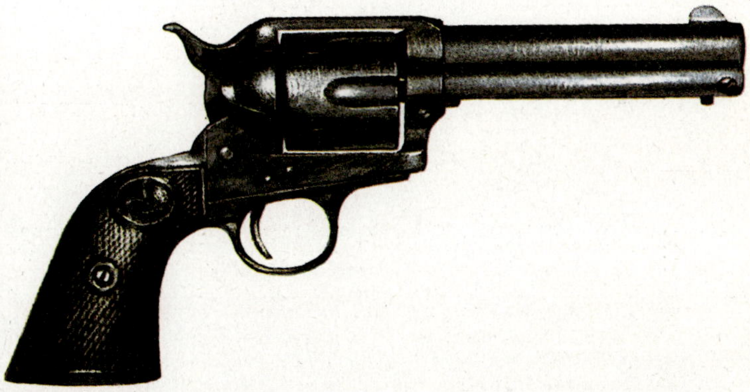

The frontier model of the Colt, .44–.40, appeared in 1878. It was the most popular of all the revolvers, and was the one used in most of the films of the West. It is interesting to note that it had the same calibre as the Winchester, and because of this both weapons used the same ammunition.

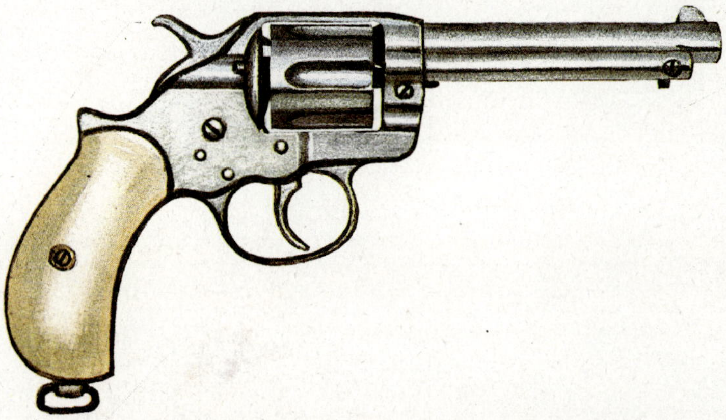

The Colt Company (Samuel Colt had died by that time) patented a double action revolver in 1878. In the drawing you can see how useful this new system was by the space that lay behind the trigger, which allowed the hammer to come all the way back when pulling the trigger.

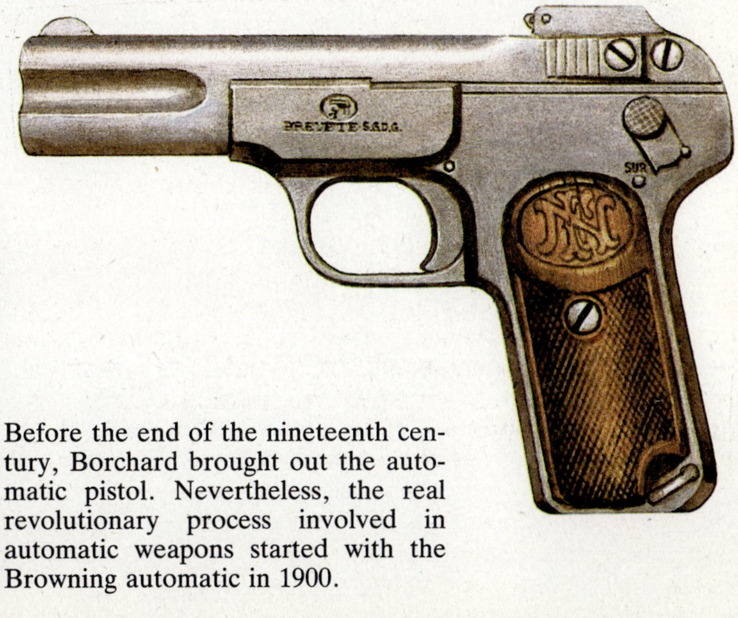

Before the end of the nineteenth century, Borchard brought out the automatic pistol. Nevertheless, the real revolutionary process involved in automatic weapons started with the Browning automatic in 1900.

THE REPEATER GUN

For many years soldiers and others who used guns had asked the designers of such weapons to produce a gun that would fire several times without reloading. The designers did not turn a deaf ear to these pleas but they were unable to come up with the answer. The revolving chamber, which was fine for small weapons, could not be adapted to rifles.

In 1848 the American inventor Walter Hunt patented a repeater rifle that was completely original. And even more revolutionary was the bullet, because it carried its own powder within it and also had a cork base. Underneath the barrel of the rifle there was a tube of ammunition and new bullets and old cartridges were inserted and extracted respectively by means of a hand-worked lever. Nevertheless, it was not completely successful until Benjamin Tyler Henry invented a metallic capsule.

Did you know . . .

. . . that Hunt's system was used by Smith and Wesson and also by the Volcania Company, which went bankrupt but then re-established itself thanks to Winchester, the manufacturer?

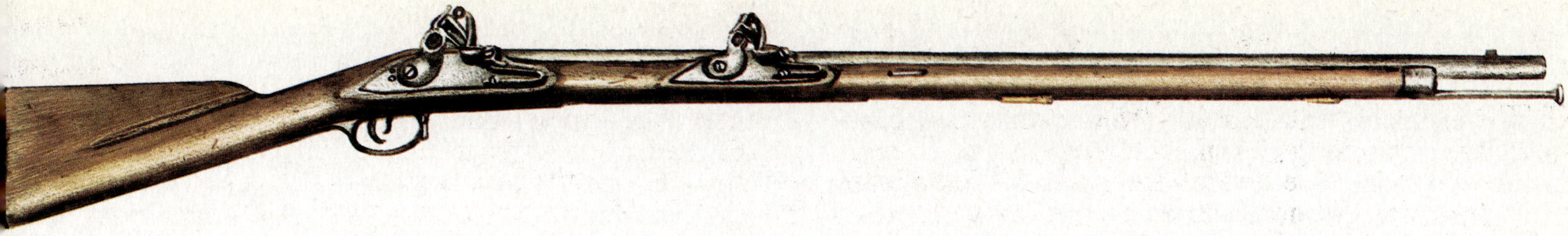

Among the many early systems, one model that deserves a mention was known as the 'firecracker'. The weapon was loaded with a given number of bullets, and there was a certain ration of gunpowder between each bullet. The main trouble with this gun was that once it started firing there was no way of stopping it until it was completely empty. The illustration shows a Chambers firecracker; it is the only one in existence.

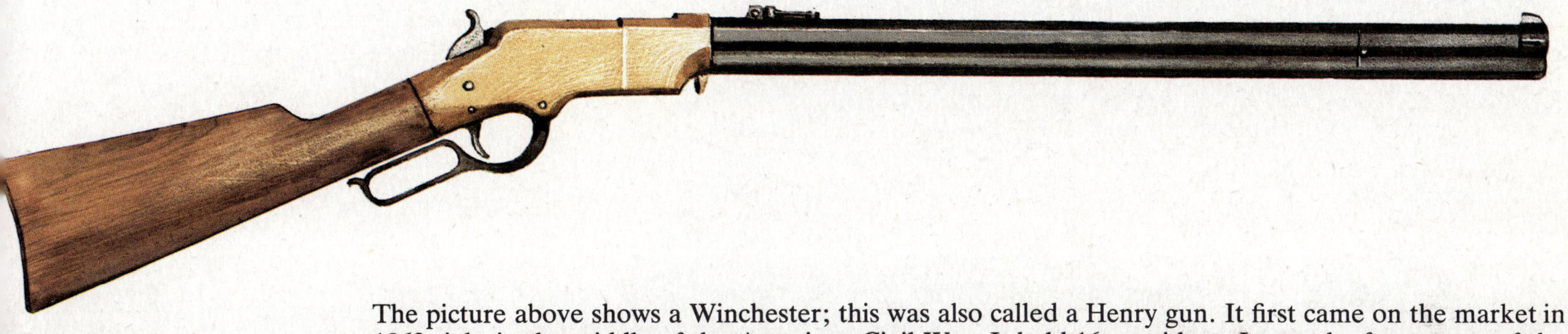

The picture above shows a Winchester; this was also called a Henry gun. It first came on the market in 1862 right in the middle of the American Civil War. It held 16 cartridges. It was the forerunner of the Winchester .66 and the .73, which became favourites of the Wild West. Strangely enough, it was rejected for use by the Northern armies, more particularly by General Wolf Ripley, who was not fond of new gadgets. It was said of this gun, as a reflection of its value: 'This gun is loaded on Sundays and fires all week'.

Above is the Spencer carbine, which was the first repeater rifle. This was also rejected by Ripley. But its inventor, Spencer, refused to take this lying down and appealed to Lincoln, who granted his request. The weapon was so successful that a few regiments used their own funds to buy it. Later the Spencer had to compete with the Winchester, and sales dropped.

The Winchester was not widely accepted in Europe. In that continent the Dreyse or the Chassepot were preferred. As a result the repeater rifle used by Europeans was a more solid affair and more powerful, but it fired more slowly.

In 1885 von Mannlicher invented the box magazine. With a simple movement all the bullets could be introduced into the gun.

The picture above shows a Mauser gun made in 1898. This was the most popular of the repeater rifles.

THE MACHINE GUN

Ever since the days of bows and arrows, machines capable of launching more than one missile at a time have been in evidence. After gunpowder had been invented, there was no lack of inventors who devoted themselves to finding a way of firing lots of bullets at a time. But in spite of all their efforts, they came up with nothing useful because the solution awaited the advent of the metal capsule. Once the metal capsule was available, machine guns of various kinds became available everywhere.

It should be noted that the first machine guns were purely manual, that is to say, the whole of the mechanism was worked by hand either by a crank or by a lever. They hardly produced the results that measured up to the effort needed to fire them, but nevertheless they were everpresent in all wars from the American Civil War until the end of the nineteenth century.

Did you know . . .

. . . that the machine gun, which was a weapon equal to at least 100 men firing individually, was called 'the devil's sprinkler', because of its dangerous nature?

. . . that the first really modern machine gun was produced by Maxim in 1884, in which all mechanical movement was activated by the power generated from the recoil?

. . . that in 1895, Browning patented a machine gun which obtained the energy for its automatic movement from just a tiny fraction of the pressure of the gases involved in the explosion?

Above you can see a machine gun worked by hand. The machine gunner used a crank. It was invented by Richard J. Gatling in 1861. It could fire between 500 and 600 bullets a minute.

On the right is a Maxim machine gun (German). It had a 7.92 mm calibre and used belts of 250 cartridges each which could be spliced together. In theory it could fire 600 rounds a minute.

In World War II various Maxim models were still in use. The recoil mechanism and breech were used by other types of guns, such as Fiat, Vickers, and Schwarzlose. The German M6-42, the best of the lot, also worked on the same principle.

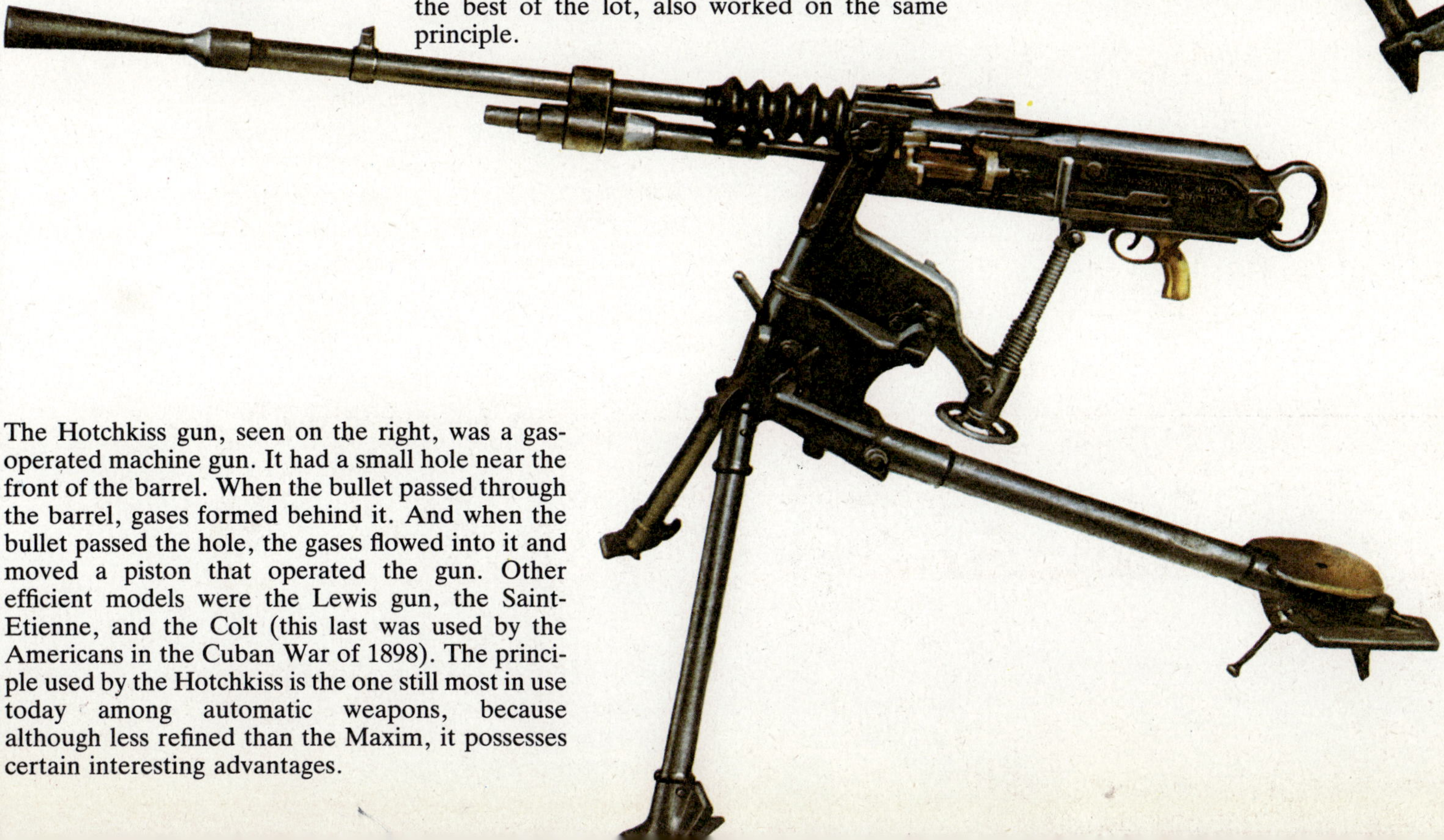

The Hotchkiss gun, seen on the right, was a gas-operated machine gun. It had a small hole near the front of the barrel. When the bullet passed through the barrel, gases formed behind it. And when the bullet passed the hole, the gases flowed into it and moved a piston that operated the gun. Other efficient models were the Lewis gun, the Saint-Etienne, and the Colt (this last was used by the Americans in the Cuban War of 1898). The principle used by the Hotchkiss is the one still most in use today among automatic weapons, because although less refined than the Maxim, it possesses certain interesting advantages.

TRENCH WARFARE

During World War I repeater weapons, especially the machine gun, forced the soldiers to adopt a new tactic to lessen casualties. This new defensive tactic was simply the digging of trenches. But any cover would serve — a shell-hole or any slight irregularity in the ground, provided it gave some protection from enemy fire. One effect of trench warfare was that both armies were bogged down and neither could advance to any appreciable extent. In spite of it all, men went on dying in a futile exercise that had no decisive effect on the battle.

Did you know . . .

. . . that out of the first primitive trenches used in World War I, there developed the formidable defensive lines known as the Maginot and the Siegfried, which nevertheless were unable to repel the enemy advances during World War II?

On the right is a German trench dug to the north of Aisne during World War I. The five numbered sections show distinct parts of the trench.

1 Forward trench for lookout.
2 Actual trench for firing from.
3 Machine gun nest.
4 Communication gallery.
5 Chamber where the props sustain the roof.

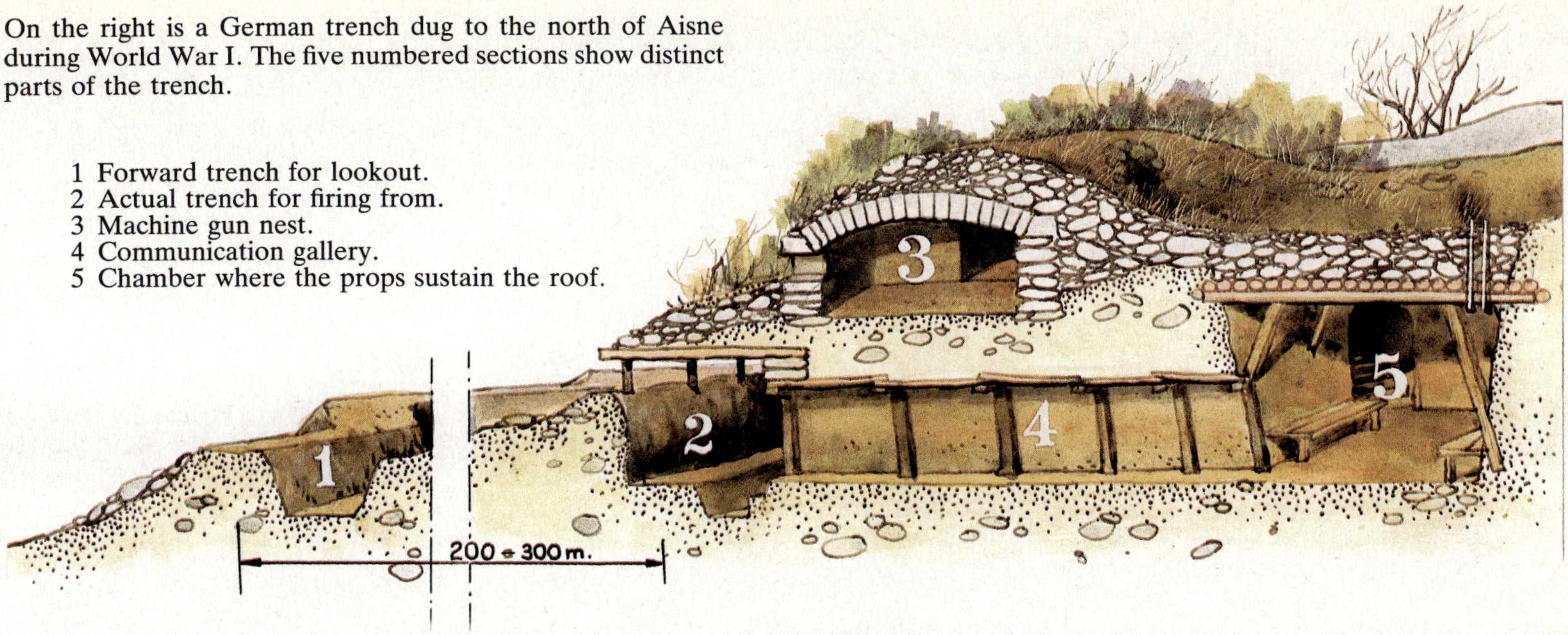

Sometimes the trenches were mere ditches dug in haste under fire. They were made in zigzag lines, as you can see from the picture below, to protect them from raking enemy crossfire.

Above and on the left are hand grenades used in World War I, and they are very similar to weapons employed in the 1940s. These weapons were vital both for attack and defence in trench warfare. Such grenades, as well as howitzers and mortars, were necessary to stop massive attacks and break up enemy formations close to defensive lines.

Below is a battle scene reconstructed from World War I. It shows an attack on the trench system. This moment of 'going over the top' was the one most feared by the troops because a man was extremely vulnerable at that instant and the act required a great deal of courage.

ARMOURED VEHICLES

On the morning of 15 September 1916, in the middle of the Battle of the Somme during World War I, the advanced German lines were being bombarded by Allied artillery. Eventually the bombardment stopped and the German soldiers braced themselves in their trenches to fight off the enemy attack which they knew would come. Nevertheless, it wasn't men they suddenly saw appear before their trenches, but machines. The astonishment and stupor of the German soldiers must have been similar to that suffered by the ancient Romans when they were first confronted with Hannibal's elephants.

Their stupor turned into panic when they found that their small arms were quite useless against these machines, which flattened everything before them. The eleven British tanks swept through the enemy lines, but they were unable to keep up their momentum and the German front held. The commander of the tanks said afterwards: 'We have made a mistake by attacking with so few units. We have given the secret away.'

And that is how the Germans came to learn of armoured vehicles and began to put into effect methods of combating them. Eventually it was discovered that small-bore cannon fire was the most effective against tanks, and this was used right up until the end of the war.

One year after this initial attack the Allies did what they should have done in the beginning: they launched an attack with a great number of tanks. But the surprise element had gone. And even as the bells of victory were ringing in anticipation in London, the Germans were already counter-attacking and destroying the Allies.

Where was the weakness of the attacks? Perhaps it was their slowness that hindered their exploiting their advantages to the full. In those days tanks were used purely to support the infantry. In World War II this was shown to be a fallacy.

Did you know . . .

. . . that when tracks first appeared on mobile vehicles, this hastened the development of the tank?

. . . that the British named an armoured vehicle 'the tank' because it was manufactured in exactly the same way as if it were a tank for holding liquid or other materials?

At the beginning of the twentieth century the idea was put into action of producing a vehicle which would be capable of resisting enemy attack and at the same time being able to attack itself. This idea had been in the minds of men for many hundreds of years. As a result of experiments, a number of models of armoured vehicles, all equipped with machine guns, were produced.

On the right is a model of a vehicle which later on was to be called a 'tank', whose armour plating varied between 3mm and 8mm. Normally it had swivelling turrets. But it had the disadvantage of not being able to cover very rough ground and, as a result, it was useless on the battlefield.

The tank shown here weighed 7 tonnes and was 5m long. It carried a crew of two. Its armour plating on the most exposed parts was 22mm thick. It was a miniature war chariot of its age. But it shared the same disadvantage as other tanks: its very slow speed.

The tank shown on the right was the German reply to the new Allied weapon. Only 20 of these tanks were made and they were not sufficient to win victory.
It weighed 30 tonnes and was 8m long and 3.2m wide. It could travel at 9km/h. It carried a 57mm gun and six or seven machine guns. It housed 18 men.

NEW DEFENCE SYSTEMS

At the same time as the attacking power of armies was being improved, it became necessary to build adequate defences. The system that in 1800 used many-angled bastions was now out of date, in view of the powerful artillery used on the battlefield. Military architects found it necessary to build exterior defences farther and farther away from the actual place being defended, in order to keep enemy cannon fire at a distance.

Because it would prove exceedingly costly to build a continuous fortification ringing a strongpoint for up to 2km the military designers decided to build instead a chain of fortifications, each one separated from the other but together covering a field of fire that would give complete protection. This idea of using small fortifications became immensely popular, and before the end of the nineteenth century they were sprouting everywhere. But the constantly increasing firepower of the cannons, and bigger and bigger explosions, brought to light certain weaknesses in this system. In 1904, the Russian forts of Port Arthur collapsed under the fire of the 11-inch howitzers of the Japanese.

Later on, the German designer Scroeter proposed building trench redoubts. These consisted of underground defences enclosed by a perimeter of barbed wire and protected by a ditch 6m deep, which could be used both for protecting the artillery and nests of machine guns, and for living quarters, built of concrete and steel. As soon as this kind of defence became common it naturally enough stimulated the attack to produce enormous guns to try and batter it down. These guns were so big that they could only be moved with great difficulty and this may have been the cause of the failure of the German offensive in 1914.

On the far left you can see a line of trenched redoubts or strongpoints. On the immediate left is a model of one of these strongpoints; and below is a section through such a fortification.

1 Wired levelled area
2 Line of small arms emplacements
3 Ditch 5m wide with barbed wire barricade
4 Line of weapons to defend the ditch and the slope beyond
5 Fortified observation post
6 Armoured artillery emplacement
7 Concrete shelter beneath the front walls
8 Ditches
9 Two-storey barracks
10 Shaft blockhouse with weapons for the defence of the ditch

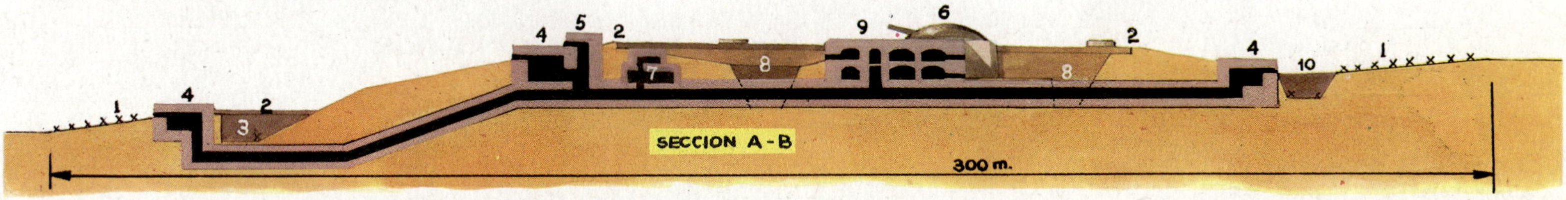

Fortified defences had to be protected from above as well, to resist howitzers and mortars. This resulted in fortified enclosures made of ferro-concrete (below).

Sometimes fortifications were equipped with naval turrets, as seen below. It should be remembered that naval artillery at the beginning of the twentieth century was already highly developed. Naval guns could be adapted perfectly to the needs of a fort and it was a practical proposition to mount such artillery taken from ancient cruisers.

In important fortifications, special turrets were mounted, protected by thick armour plate. The turret and the gun inside it were both retractable, that is to say both before and after firing they were housed out of sight. When the French troops in 1916 were able to re-take these strongpoints at Verdun they could see that the German shells had done little damage to the armour plate, although they were not able to verify this in other cases because the forts were taken so easily by the Germans.

ARTILLERY RULES THE ROOST

Just as automatic weapons were steadily progressing, so did advances in artillery keep step. By the beginning of the twentieth century, the artillery had reached a high state of efficiency. Its peak was achieved during World War I. Later, aerial bombardment and guided missiles put an end to its practical functions. But until that time the artillery was a deciding influence in all wars that broke out during the twentieth century, and was even used at the beginning of World War II.

Breech-loading and rifling improved the big guns, but the real improvement came with the invention of the hydraulic brake, which absorbed the recoil of the barrel by means of hydraulic shock-absorbers, and which also placed the firing cube over the gun carriage after each discharge. All this helped to speed up the rate of fire and increase the accuracy of the gun. One of the improved models was the 75mm Schneider. Eventually all the great powers equipped their guns with the hydraulic brake.

Did you know . . .

. . . that the Germans, who were always fascinated by anything extremely large, endeavoured to build the biggest and the heaviest cannons in the world, like the one illustrated here, which was built by Krupp and called Bertha in honour of his daughter?

The most sensational of the German big guns in World War I was the so-called Paris Cannon, shown on the right. It was christened by the French 'Big Bertha'. It was a 'Long Max' naval cannon with a calibre of 381mm, and it was mounted on a railway waggon. It was accompanied by another cannon of 210mm calibre with a barrel length of 35m. This super cannon enabled the Germans to bombard Paris from a distance of 120km. It was a sensational development for those days. The Germans produced about 12 of these giant guns, which showered Paris with between 400 and 800 shells, and caused hundreds of deaths and 1,000 wounded.

The artillery used during World War I was made up of mortars, howitzers and cannons, which was the mixture as before, but these weapons were by now much more efficient. On the left you can see a Skoda mortar with a calibre of 305mm, on which the recoil cylinders mounted on the barrel are conspicuous. Below is a picture of a 380mm howitzer, also manufactured by Skoda, used in 1916. It had a range of 15km.

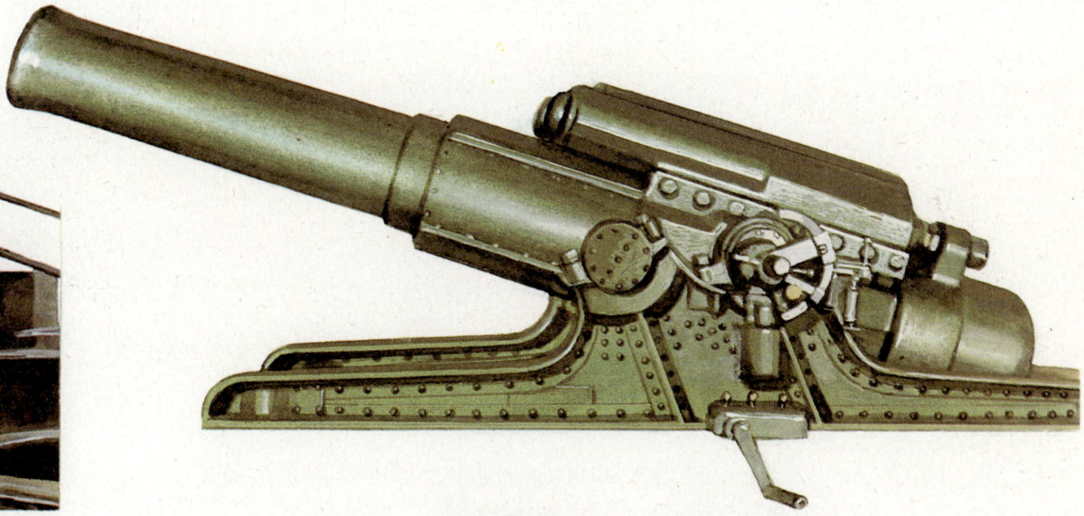

On the right is the German version of the famous French 75mm field gun, although this model is a shortened version of 24mm calibre. With this gun the Germans were able to take on tanks with great success. The gun could accommodate six different types of ammunition. Its value was fully justified in the Battle of Villers-Cotterets, in July 1918, when it wiped out almost half the enemy tanks.

AERIAL OBSERVATION AND BOMBING

For many centuries military tacticians had been dying to find a way by which they could discover what the enemy was doing behind his defensive lines. But such observation would have to cover a great deal of ground, and this meant that it would have to be done from a considerable height. And this is what the first manned balloons were used for. In World War I aeroplanes, airships, and balloons were equipped with the finest telescopes and the best available photographic equipment.

But with the knowledge that one could fly over enemy territory came the idea that one could also drop bombs on it. By 1910 there was already in existence a Michelin competition for accuracy in aerial bombing. Nevertheless, the only aerial vehicle that was big enough to carry a practical load of bombs in 1914 was the airship. However, the huge bulk of this craft was a serious disadvantage.

Germany used the airship more than any other country in World War I, and by its means she began a systematic campaign of bombing. But by 1915 casualties among airships had become so great that they were relegated to occasional night flights. This led to the idea of using aircraft for bombing the enemy lines.

Did you know . . .

. . . That in the beginning the role of the aeroplane was purely that of reconnaissance, and that if two machines from opposing forces crossed each other's paths the occupants saluted each other?

. . . that a little later, one pilot decided to react with his pistol instead of with his hand, and from then on the fight was on?

Captive balloons were extremely useful for observation. When they were knocked down, the pilot was usually able to bale out, using a parachute.

After the summer of 1914, reconnaissance of the enemy rear lines was done only by aeroplanes. Balloons were used less and less. But aeroplanes limited themselves only to reconnaissance, and had no intention of attacking the enemy.

After a relatively peaceful interval, all kinds of missiles began to be hurled from the aeroplanes. Among them were the curious arrows shown above, which could transfix a man. They were made of steel, were 30cm long, and they were launched in groups of 100.

In 1915 aeroplanes became specialized, and some were built purely to bomb enemy targets. The bombs were thrown over the side by hand or launched by means of very primitive tubes. Little by little these launching gadgets were perfected. By 1917 there were already bombers in the sky which could carry 1,300kg of bombs each, at a speed of 140km/h for a distance of 500km.

The heaviest bombs were launched by the Germans, and they weighed 1,000kg each. it has been estimated that the Germans launched about 27,000 bombs during World War I with an average weight of 25kg each. On the right you can see some of the bombs that were launched; and below is a picture of a German Gotha.

BOMBS USED IN 1916

THE GOTHA G5, A 1917 GERMAN BOMBER

THE FIGHTER PLANE

Aircraft had already begun to reconnoitre enemy territory, and had been doing this with considerable success. For example, on 30 August 1914 the Germans were able to halt a Russian advance by means of intelligence received from aerial reconnaissance. And in September of the same year French aerial reconnaissance discovered some of Von Kluck's army movements and this enabled them to take up defensive positions for the forthcoming Battle of the Marne.

In view of all this, the combatants decided that it was vital to be able to destroy these spy aircraft. And what better to do this than another aeroplane which was to be called a fighter plane, and which would be armed with machine guns? the fighter plane had to be faster and more manoeuvrable than the reconnaissance and bomber aircraft that it opposed. And it also had to be much lighter.

Did you know . . .

. . . that at first pilots used to carry pistols and side arms to fire at the enemy pilots, and they even threw grenades at them?

Pistols and other small arms were the first weapons that were used in aerial combat. In the picture below you can see a German observer using an automatic Mauser pistol.

The ideal weapon was the machine gun. In the picture below an infantry soldier is firing a machine gun over the top of the aircraft's propeller.

In 1915, the Germans claimed that the famous pilot Roland Garros aimed by pointing his aircraft at the enemy planes. It is certain that Garros was the first man to fire through the propeller, which was protected by an armour-plated deflector.

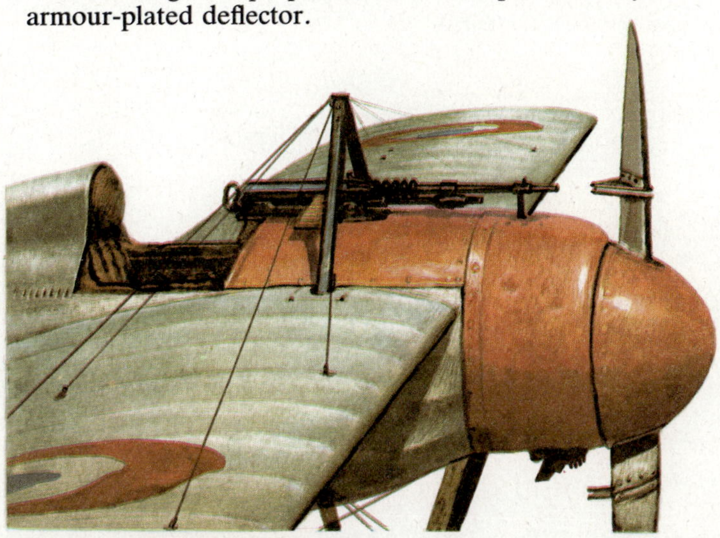

Because of a navigational error, Garros was forced to land behind the German lines. They discovered his firing secret and managed to improve this device on their Fokker aircraft. But by mistake a Fokker landed behind the Allied lines and they copied the improved version.

Parachutes were hardly used during World War I. Some nations banned their use, saying that if a pilot were equipped with one he would not try to save his aircraft before baling out.

The simple act of flying in those days was a feat of great bravery. As a result, there grew up between fliers a kind of medieval chivalry which showed itself in the respects paid to fallen enemy fliers.

THE FIRST ANTI-AIRCRAFT GUNS

In 1914 aviation was not regarded as anything to be worried about. As a result, there was no anti-aircraft artillery. Nevertheless, Germany was aware of the dangerous possibilities of aircraft, and by 1910 had already manufactured various prototypes of anti-aircraft guns in order not to be caught napping. At first reconnaissance aircraft flew very low and, as a result, were in range of hand guns. But when they flew a bit higher, field guns began to be used against them. These had to be equipped with gun carriages that could swivel through 360°, time fuses, and special range finders in order to make sure of hitting the target. All these were necessary in order to make them worthwhile under battle conditions, but once they were in use, anti-aircraft guns seemed set for a fruitful future.

In 1915 two 76mm anti-aircraft guns were mounted on the battleship *Queen Elizabeth*. Soon other guns, which were built specially for anti-aircraft duties, were being manufactured. They were long and narrow with a great range and small calibre, mounted on carriages that would allow them to be elevated almost to the vertical. Nevertheless, during World War I more aircraft were shot down by fighters than by anti-aircraft weapons.

CAMOUFLAGE AND IDENTIFICATION

It should be emphasized that the war of 1914 was the most revolutionary of all wars, because during the course of the conflict dozens of traditional ways of thinking were overturned and there dawned a new era in warfare. In spite of its great technical innovations and inventions, World War II was merely a continuation of World War I in many respects. In other words, in 1939 everything that had been rather primitive and clumsy in 1914 was merely improved and perfected.

An example of improving ideas that were originated earlier can be seen in the use of camouflage, which means hiding or disguising something from the enemy. It was first used in World War I. At that time whole towns were already being painted to disguise them. Nets were placed over big guns to hide them, and aircraft were painted the colour of the ground above and the colour of the sky below.

And what about the aircraft? They all looked so much alike that a system of identification had to be put in hand. In order to do this national markings were used, and this system is still in use today.

Did you know . . .

. . . that the unfortunate pilots of 1914 were faced with being fired at by the enemy on their outward mission and coming under fire once again as they crossed their own lines because all aircraft looked so much alike?

. . . that during a dogfight in World War I a British pilot who had got lost among the clouds decided to land when he saw some French soldiers beneath him, but found out to his dismay that those soldiers happened to be prisoners of the Germans?

COMMUNICATIONS AND SIGNALS

Military information became extremely important in World War I. It was no longer sufficient to have a bugler sounding the command nor for the postman to distribute the news. Something better was needed to transmit messages over hundreds of kilometres. As a result, new tactics were evolved for transmitting information. Visual signalling came into its own and aircraft were equipped with cameras and radios to give immediate news of troop movements. Observation posts were also used, as well as carrier pigeons and messengers on horseback, on bicycles, and on motorized transport.

Did you know . . .

. . . that the new science of camouflage was able to work wonders in siting observation posts, putting a false tree where a real tree was but making it hollow, armour-plated and equipped with a telephone?

. . . that the Eiffel Tower was used as a gigantic aerial in an attempt to capture the transmitted wireless orders of the German High Command?

. . . that in order to transmit orders, all kinds of methods were used, including radio, telephone, and visual means such as reflecting mirrors and flashlights?

. . . that the soldiers who served as messengers sometimes performed heroically as they carried out their duties under heavy fire?

THE RETURN OF THE FLAME-THROWER

In February of 1915 parts of the French front near Malancourt were taken by surprise with a baptism of fire. For the first time in this great conflict flame-throwers were used. This 'new' weapon was no more than the old 'Greek fire' brought up to date. It was particularly effective against trenches, fixed positions, machine gun nests, and fortified places.

There were two kinds of flame-throwers: the individual light machine, and the heavy equipment which was carried on a vehicle. Both kinds carried tanks — one large, with some inflammable liquid, usually petrol; and one small, carrying some inert gas, usually nitrogen under pressure. The liquid came out of a metal tube called the 'thrower', which was connected to the large tank by means of a rubber hose.

The flames were up to 20m long. Nothing could live in the intense heat throughout a zone 30m in depth and 15m wide for each flame-thrower.

THE HORRORS OF GAS

The date was April 1915; the location, the Ypres front. It was four o'clock in the afternoon. Soon the expected north wind began to blow. The German sappers began to open a number of bottles with great care. A greenish yellow cloud, which later turned whitish, began to sweep towards the enemy lines. The scene was terrible. There were many victims of this poisonous gas. Nevertheless, the Germans made the same mistake with gas as the British had with their tanks. They failed to press home their advantage, because they had not expected such a great success. In merely a matter of days, gas masks were being worn on all fronts, and gas attacks no longer caused surprise nor panic, and presumably casualties dropped. From then on began the era of the so-called choking gases.

The reason that gas was not used in World War II was not out of any sympathy for the victims, but out of fear. All armies involved in that war possessed poisonous gas, and they also had gas masks, and the danger was very real for everybody. But what happened was exactly what occurred with the atomic bomb — without a monopoly of the weapon it became a deterrent, because nobody dared use it first.

Did you know . . .

. . . that in 1918 corporal Adolf Hitler was burnt by mustard gas (gas which produces blindness, burns on the skin, and ultimately results in death)?

. . . that five tablespoons well distributed of the ester (a chemical compound) of phosphoric acid would be enough to paralyse a whole city?

SPYING: INFORMATION FROM UNDER COVER

Information and spying are close bed-fellows. Any national information service should be in a position to know what is going on both inside and outside the country. And in order to do this there must be spies.

Spying is as old as man himself because spying entails lying in wait, using cunning, craftiness, and subtlety. One simple piece of information obtained through a spy can change the whole course of a battle. All nations use this kind of service and in wartime it becomes extremely important, although a spy discovered on enemy territory is doomed.

In World War I the spy became known as more than a simple traitor. Some spies were legendary: it is enough to remember Mata Hari, the famous woman spy. In World War II, also, there were plenty of clever spies, including Sorge, the Cat, and a whole host of anonymous personalities. And in the postwar period in our day there are any number of spies emulating James Bond, sometimes acting as double agents and thinking only of the money they can earn.

The aircraft above is a Junkers JU87 Stuka dive-bomber, which was used by the Germans in Spain during the Civil War. It was terrifying looking and often had a siren attached to it, to make it sound more frightening when it dived.

A TESTING GROUND FOR WORLD WAR II

The period between the end of World War I and the beginning of World War II was one of unthinking optimism by the victorious Allied powers, which dreamed of eternal peace. Nations and governments wanted no more wars. But as soon as Hitler came to power, German war plans began to develop. The Allies began to rearm under a distinct disadvantage because they had not tried out their weapons in local wars as had the Russians and Japanese in China, the Italians in Ethiopia, and the Germans, Italians and Russians in Spain. The Spanish Civil War enabled the Germans to try out their aircraft and armoured vehicles for use against their eventual enemies, the Russians.

The picture above shows the German medium bomber, the Heinkel HE 111, which was used regularly by the Condor Legion in Spain during the Civil War from 1936 to 1939. At that time that aircraft could travel at 400km/h at a height of 4,000m. It could outdistance most enemy fighters.

Below is a picture of a Messerschmitt 109, model B, used by the Condor Legion. This aircraft with its great speed caused a sensation when it first appeared, although its firepower was not very great and it was not particularly good at manoeuvring.

On the right is a Russian fighter, the Polikarpov 1-16, which was used by the Russians in Mongolia. It was the first low wing monoplane fighter with retractable undercarriage used by the Russians. It was a better aeroplane than the German Messerschmitt 109. The Russians used almost 500 of these machines in the Spanish Civil War. They actually manufactured 20,000 of them. It was also used in great numbers at the beginning of World War II. It had many nicknames. In Spain it was known as 'the Rat' and 'the Fly'. The Russians called it *Thsk*, which means 'little donkey'.

As far as armoured vehicles were concerned, the Spanish Republicans had better models than the Nationalists, although they were not used with such efficiency. Franco's army had ancient Renault vehicles, Italian Fiat and Saldo armoured cars, and German PKIs. The Republicans had Russian T26s and BTs. In the picture on the left is a Russian BT. It weighted 12 tonnes and could travel at 86km/h on wheels and 62km/h on tracks.

ADOLF HITLER

Germany lost World War I and the postwar years were for her bitter and disastrous ones both socially and economically. But a man called Hitler, born in Austria and who had served as a corporal during the war, believed that Germany had not really been beaten. This idea drove Hitler into politics. He built round him a large group of fanatical followers and with them formed a political party — the Nazi Party (National Socialist Party). He took part in a number of street riots and was duly imprisoned. While he was in prison he wrote a book called *Mein Kampf* (My Struggle), which outlined a programme for his party.

After a number of victories and defeats the Nazi Party managed to obtain representation in the Reichstag (German parliament) in 1933. They put aside legal considerations to seize power, overthrowing the opposition and capturing control of the army and the law. Finally Hitler proclaimed himself *Fuhrer* (dictator). With his nation once more back under law and order and the army rearmed, he decided to reoccupy the demilitarized Rhineland region. Later, Hitler occupied the Sudeten region of Czechoslovakia, occupied Austria, and the rest of Czechoslovakia, and finally went to war over Poland.

MISPLACED FAITH IN DEFENCE

World War II began as a fighting war where the first one left off, that is to say as far as technical improvements to weapons were concerned. In battle strategy the French in particular were convinced that this was going to be another war of attrition (wearing away), like the first one. As a result, they put all their faith in the Maginot Line. The Germans also had a defensive line called the Siegfried Line. The Maginot Line was started in 1927 when André Maginot was minister of war. The line was made up of fortifications which were nearly all underground. On the surface there were ramparts of steel and cement, ditches and anti-tank obstacles, barbed wire, and minefields. But the Germans outflanked it when they invaded Belgium and the line was never tested.

1 Observation post. 2 Concrete strongpoints with steel cowlings. 3 Ventilator shaft. 4 Air conditioner. 5 Barracks. 6 General stores. 7 Signals and administration offices. 8 Lift to gun emplacements. 9 Subterranean munitions railway. 10 Generators. 11 Magazine. 12 Troops' quarters. 13 Sick bay. 14 Medical store and grocery. 15 Goods lift. 16 Toilets and washrooms.

BLITZKRIEG: THE LIGHTNING WAR

The value of tank warfare was discovered in Poland by the Germans, although they used these machines quite differently from the way they were used in World War I. The new method of using tanks was based on speed and powerful attacks by massed tanks. The enemy front was attacked with a number of tanks or other armoured vehicles, a breach was made in the line, and through this motorized infantry poured like a river. The infantry penetrated as deeply as it could and then proceeded to fan out and surround the enemy forces, attacking them from the rear and cutting off their supplies. The blitzkrieg also needed the co-operation of the airforce.

This new method of lightning warfare was later tried out in France. The French and the British had more tanks than the Germans, but they used them as aids for the infantry, and did not group them in individual divisions. The Nazi armoured columns advanced in mass formation and flattened everything in their path, although at times the German commanders were a little worried that their enemies might be able to use tanks in counter-attacks and cut off their advanced elements. At first this did not happen, although later it was used successfully by the Allies.

Did you know . . .

. . . that the people most surprised at the success of the German armoured columns were the Germans themselves?

. . . that it was a French general, Charles De Gaulle, who was the first to write about new tank tactics and the French took no notice whatever of his predictions?

German tank, PZKW Mark IV

We have already noted that the Allies had more tanks than the Germans, but the German vehicles were faster and better built. This was especially true of the size of the gun turret, in which they could fit two or three men. The allied turrets could only take one, and this was quite inadequate because he had to load the gun, aim it, and also command the vehicle and transmit the orders.

The best German tank of World War II was the PZKW Mark IV, which is shown on page 36. Only 278 of these tanks were ever built. The picture below left is the French CHARB tank, armed with a 75mm cannon below and another 47mm cannon in the turret. Below is a picture of a Matilda tank, a British model which was protected by 70mm armour plate.

On the right is a Russian T34 tank which produced admiration and fear among the Germans; they thought their own Panzer 3 and 4 models were the best in the world. Until 1943 the T34 was the best tank in the war. It had angled exterior armour plate; a 76.2mm cannon; a speed of 50km/h; and an engine that ran on non-inflammable diesel fuel, which gave it a range of no less than 400km. The Russians produced 10,000 of these tanks a year. It is no wonder that eventually the Germans were defeated by the Russians on Russian soil.

In order to compete with the T34 the Germans produced new tanks. One of these was the PZKW Mark V, the Panther (shown on the left), which appeared in 1943 and had armour plating that was 100mm thick and carried a 75mm cannon. This gun could penetrate up to 140mm through steel plates one kilometre away.

HITTING BACK AT THE TANKS

'Fire!' In the interior of the turret of a Mark I Tiger tank the explosion, the smoke, and the fierce recoil of the 88mm cannon blended with the terrible racket of the tank itself. The anti-tank missile had hit a T34 in the side and had smashed its tracks. There was a tremendous explosion and the Russian tank's turret sailed into the air for some 20m. This was proof that the best way of fighting tanks was to oppose them with other tanks. The decisive weapon here was the power of the gun. One of the great influences on the fortunes of anti-tank battles was the behaviour of the tank crews themselves: they had to be swift and adaptable as well as courageous. Armoured vehicles were quite defenceless against aerial attack, so if an army lost command of the air this was fatal for its armoured columns.

Anti-tank guns date from World War I but they were only really efficient from 1939 onwards. They were easily recognizable by their long, narrow silhouette. The gradual growth in the size of tanks in general meant that the calibre of such guns had also to be increased.

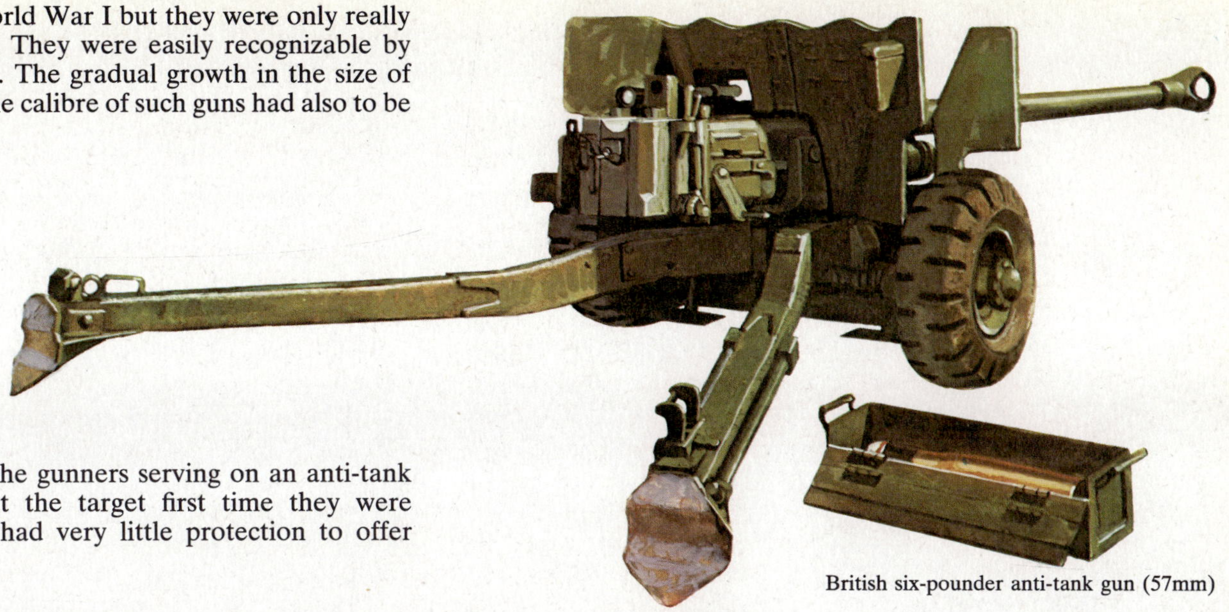

There was great danger for the gunners serving on an anti-tank gun because unless they hit the target first time they were exposed to enemy fire and had very little protection to offer against it.

British six-pounder anti-tank gun (57mm)

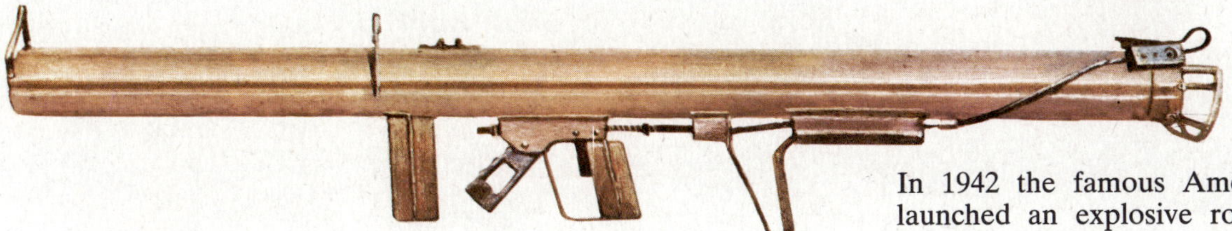

In 1942 the famous American bazooka gun appeared. This launched an explosive rocket. The Germans had their own version, called the Panzershrek, seen above.

Anti-tank mines were in general use: they were buried in the earth and exploded on contact. In this way whole minefields were built. But their value was more as a deterrent, because once the first tank had been blown up by a mine the others prudently refused to advance any farther. In the picture below you can see a German Teller 43 mine.

The magnetic mine (below) was more effective than the ordinary mine (below left), although more risky to use. The operator hid in a ditch until a tank passed by and then attached the mine to the tank. The mine had a time fuse and exploded within three or four minutes.

Tanks were vulnerable to certain obstacles. Because of their limited visibility they could often fall into traps. Fortifications usually had anti-tank barriers in front of them, and these were located at the bottom of ditches or moats with concrete pylons called 'Dragon's Teeth'.

During World War II there appeared a number of tank hunters. These were armoured vehicles that lacked turrets but mounted anti-tank guns. The one shown on the right is the Nazi Elephant model. It weighed 68 tonnes, had armour 200mm thick and was equipped with the famous 88/56 gun, which could penetrate steel plate 141mm thick.

BOMBS TAKE THE PLACE OF ARTILLERY

By 1939 the airforces of the warring nations had developed greatly, whereas the artillery had remained relatively stable. Nevertheless, the Germans had built a huge cannon called Dora, and this was the swansong of the artillery. It weighed 1,000 tonnes, had a barrel 27m long, and fired 800mm shells to a maximum distance of 471km. It was obvious by now that for long-distance bombardment aircraft would replace artillery.

The Nazis had no heavy long-distance bombers, but the British had. In 1942 they produced the four-engined Lancaster bomber, which could carry more than 10 tonnes of bombs and had a range of 2,700km. The Americans also built a number of different kinds of heavy bombers during the war. These four-engined aircraft could roam deep into enemy territory and reach factories, communication centres and cities deep inside Germany. The Allies used their bombers in massive raids over Germany, reaching vital objectives and depriving the Nazis of fuel. Almost at the end of the war, Dresden suffered an incendiary raid by night that caused 200,000 deaths. The mistake the Nazis made in not building heavy bombers cost them dear.

When bombing stationary objects of any size, aircraft dropped their loads from great heights, but when precision bombing was required, such as against a bridge or a ship, they had to come down low. The divebomber (left) was a special aircraft because it was equipped with brakes which enabled it to come out of its dive at the right moment.

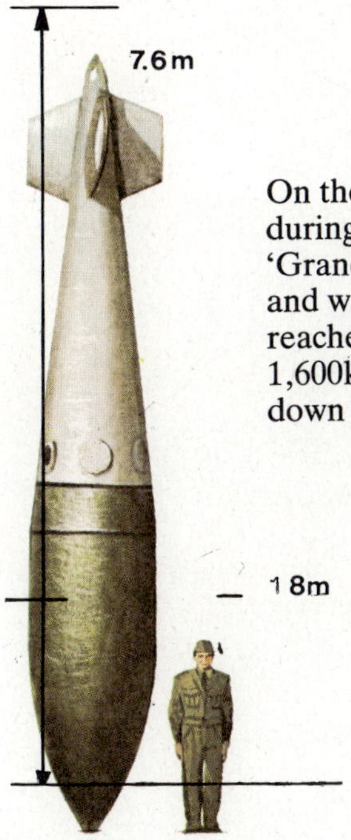

On the left is the biggest bomb used during World War II. It was called 'Grand Slam', weighed 10.5 tonnes, and was carried by the Lancaster. It reached the ground at a speed of 1,600km/h and dug into the earth down to a depth of 30m.

A curious system of flying artillery appeared when an anti-tank gun was mounted on an aircraft. The Henschal 129 B3 was equipped with a 75mm cannon, seen below, and this caused many casualties in the Russian heavy armoured columns. In case of emergency, the gun could be ditched. This kind of weapon, however, soon gave way to airborne rockets.

The greatest danger to the bombers was attack by enemy fighters, and that is why they carried their own defence. The heaviest bombers were truly flying fortresses. The Liberator, shown below, carried ten half-inch machine guns which could fire in all directions. If one of these bombers were to break formation, it would almost certainly fall prey to enemy fighters.

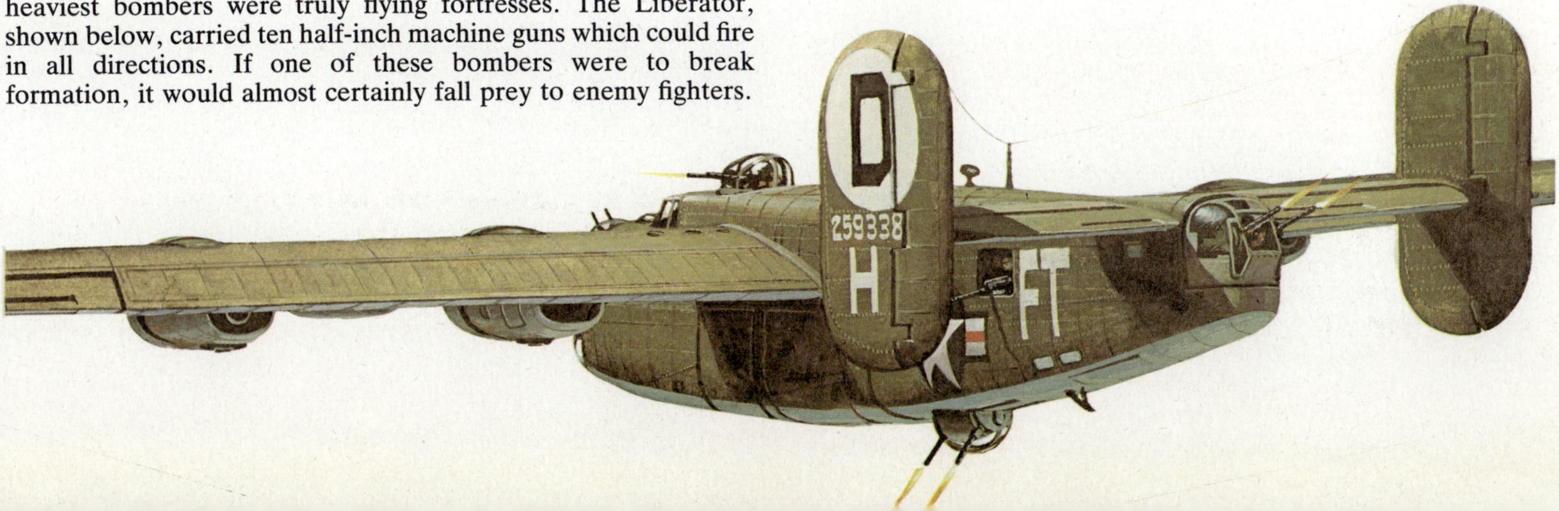

FIGHTER VERSUS BOMBER

The Liberator flew into the German pilot's sights. The fighter was for the moment beyond the range of the bomber's defensive armament, but the fighter's 20mm cannon, helped by the speed of the aircraft, found their target.

The whole fighter shook as it poured withering fire into the four-engined aircraft but, as it overshot, the ME 109 was hit several times by .5 machine guns firing from the enemy's rear turret. The German pilot must have realized that his victim was no pushover.

In the 1940s the development of military aviation already foreshadowed the possibilities that have been achieved today with bomber aircraft. In less than ten years aircraft passed from the biplane to the jet, from 400km/h to 1,000km/h, and from the piston engine to the rocket.

Above is the legendary Spitfire fighter, which was the main architect of victory in the Battle of Britain. It looked rather like the German ME 109, but it was more manoeuvrable, although its climbing speed was inferior. The early Spitfires were not so good in a dive but later models overcame this difficulty.

The Japanese had some magnificent aircraft. In spite of Allied scorn, Japanese fighters were better than American fighters, at least during the first part of the war. The Zero fighter, shown above, was faster, more manoeuvrable, and had a greater range than any American aircraft. At the end of 1943 the Japanese designed a fighter called the Hayate K84, which was believed to be the best of its kind used in the Pacific.

The best American fighter was the P51 Mustang, shown above. Its maximum speed was 704km/h at a height of 7,000m. Its range was 1,500km. Mustangs were used on all fronts. They claimed to have destroyed some 4,000 enemy aircraft in Europe alone. In *The Aeroplane* magazine a reviewer wrote: 'The pilots are so enthusiastic about this aircraft that they run out of words of praise before they come to the end of the qualities that they are praising'.

Because the Germans had some startling and easy successes in the earlier part of the war, they more or less abandoned their research into new aircraft design, but later, after suffering some defeats, they renewed their research. As a result, they came up with some formidable weapons, although unfortunately for them, these came too late to save them. Among conventional aircraft used during the last few days of the war was the Focke Wulf TA 152C. It had a top speed of 740km/h at a height of 10,300m. Its range was some 1,200km.

The most revolutionary of all World War II fighters was the Messerschmitt 262, shown left. It had arrow-shaped wings, multiple ailerons, controls worked by servomotors, ejection seat, radar, air-to-air missiles, and other advanced features. It had a top speed of 780km/h at a height of 6,100m. Its armament consisted of four 30mm cannon and 24 R4M air-to-air missiles. By the war's end, the British also possessed a jet fighter — the Gloster Meteor, although it was not as good as its German equivalent.

ANTI-AIRCRAFT DEFENCES

In addition to fighters, bombers also faced other enemies which could not be ignored. These were anti-aircraft guns firing from the ground. Such guns at first had a calibre of 75mm, but as the bombers flew higher and higher, so did the calibre of the guns increase. The heavy guns did not try to hit a specific target, but instead formed a compact barrage of explosions at the height at which the enemy aircraft were flying, and hoped by this means to destroy them either with a direct hit or a near miss.

Anti-aircraft machine guns were only useful against low-flying aircraft, and even then they had their problems. These were due to a shortage of tracer bullets, the lack of any explosion on contact, and the difficulties of trying to latch onto an aircraft that was flying past at such a speed that the gunners barely had time to adjust their sights.

Did you know . . .

. . . that thanks to radar the British won the battle against the German airfleets?

. . . that in 1943, during the bombing of Hamburg, the Allied aircraft threw down tonnes of aluminium foil to confuse the German radar?

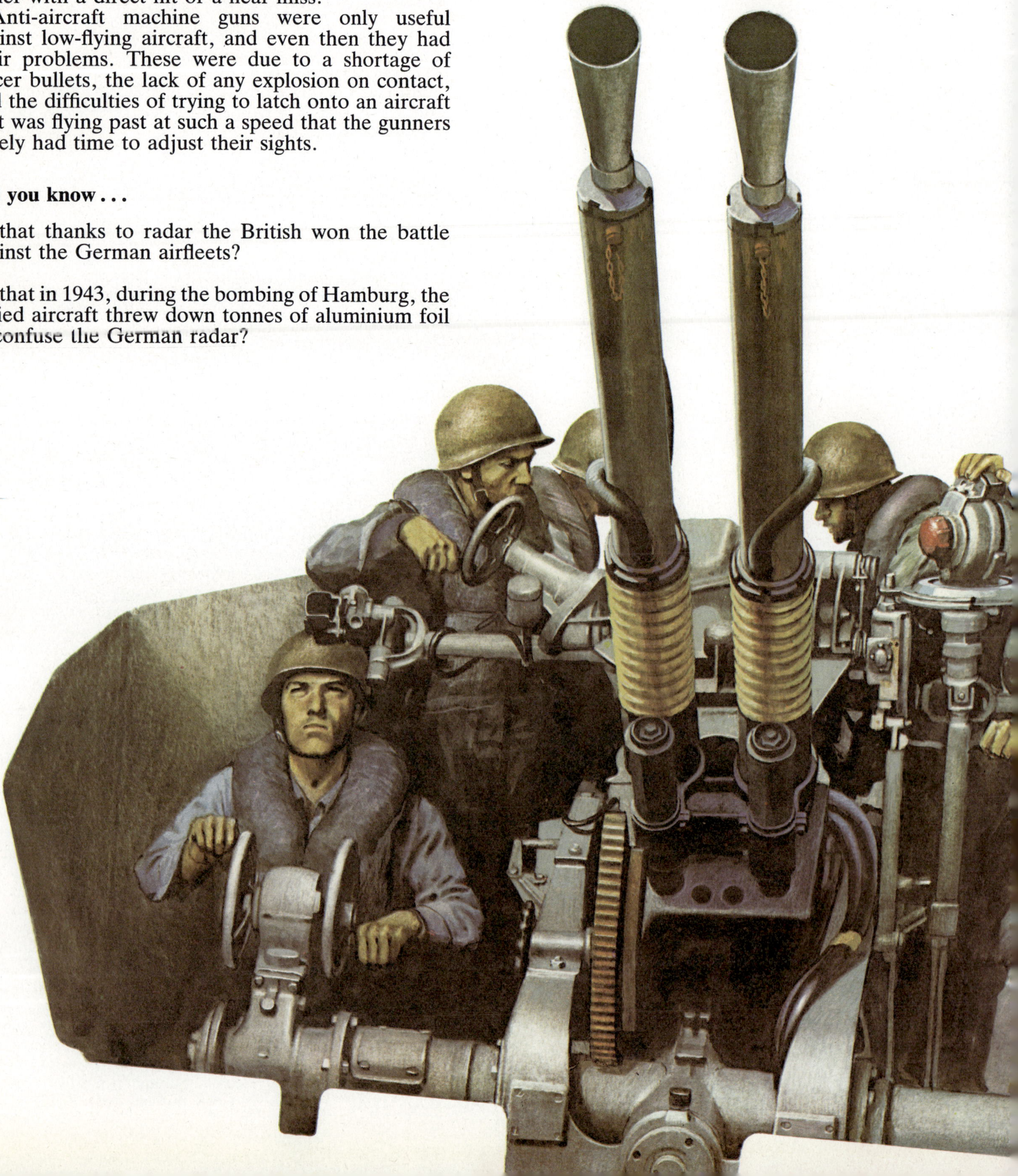

In the picture on the left you can see a German anti-aircraft gun with a calibre of 105mm. This was a version of the legendary 88 with a maximum range of 12.8km.
Another defensive measure used against enemy aircraft was the barrage balloon (below left). This was employed by the British and it proved to be extremely effective, especially at night, because the cable that attached it to the ground was quite able to cut an aircraft in half on impact. Attacking aircraft were painted black and had to be picked up by searchlights in order to be spotted.

The British were the first to perfect and use radar to detect the presence of enemy aircraft. It was because of radar that they defeated Goering's air armada. The Germans, however, soon caught on, and by the middle of the war were using radar in their night fighters, with their anti-aircraft guns, and with their searchlights.

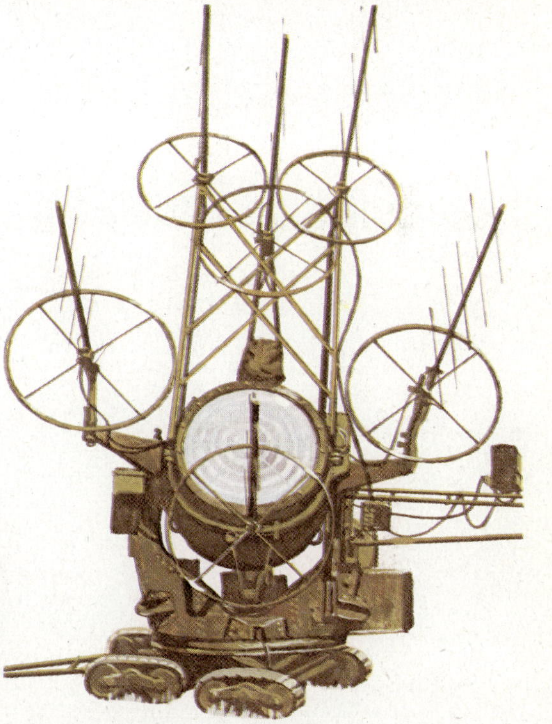

Today there are many ways of combating radar. Some of these include electronic interference, special paintwork that produces false echoes, gadgets that distort the waves, and other systems, many of them on the secret list. During World War II radar was so new that there was no real counter-measure used against it. But the British did use aluminium foil to confuse the German radar, although the Germans soon learnt to distinguish between false echoes and the true echo produced by an aircraft.

REPRISALS FROM SPACE

At half-past-six in the evening of 8 September 1944, in London, there was a sudden strange screaming noise which came just a fraction of a second before a terrible and prolonged explosion. Suddenly a dozen houses completely disappeared. There had been no air raid warning and there had been no enemy aircraft overhead. Nor had the distinctive sound of the V1 flying bomb been heard. What had happened was that the British had become the first victims of the fantastic V2 rocket. The Germans launched 1,500 of these missiles at England and on the port of Antwerp, in Belgium. But, like all the German secret weapons, it came too late to be effective. General Eisenhower has been quoted as follows: 'If the Germans had been able to perfect and use their new weapons before the invasion took place, that operation would have been extremely difficult, if not impossible, to carry out.'

The V2 was the first long-range missile in history. It was the forerunner of our spaceships and all the current nuclear missiles. It was 14m high and had a diameter of 65cm. At its launch it weighed more than 12 tonnes and it carried a tonne of high explosive in its nose cone. And there was absolutely no way of intercepting it.

On the left is a V1, the first of the flying bombs used by the Germans. It was a very simple jet aircraft whose engine cut out at a given distance and it thereupon dived to the ground with its 1,000kg of explosive. The Germans fired 9,300 of these flying bombs at targets in England but fewer than 6,000 found their mark. The number launched in all operations was about 20,000.

The British used fast fighter aircraft to attack the V1s, but it was dangerous to get within 200m of them because if they exploded in the air the fighter was also destroyed. Some daring pilots flew close to the flying bombs and with their wing tips knocked them off course.

The Allies were astonished when they found out about the plans and new designs for rocket weaponry in an advanced stage of manufacture by the German technicians. A few of these weapons were actually used during the war, but most of them never got past the experimental stage. Nevertheless, they became useful prototypes for a good many of today's modern weapons. The Komet, for example, shown above, was a revolutionary fighter that could fly at almost 1,000km/h. It carried two 30mm cannon and eight 5cm rocket-launchers above the wings or 24 below the wings. Fritz X, shown below, was a 1,500kg bomb. It was released by a bomber and was guided by radio to its target. One of these bombs sank the battleship *Roma.* On the right is a picture of the Rheintochter ground-to-air missile. It was a two-stage rocket with a proximity fuse, and it never passed the experimental stage. The Germans were also at work on a gigantic missile to launch against the United States from Europe.

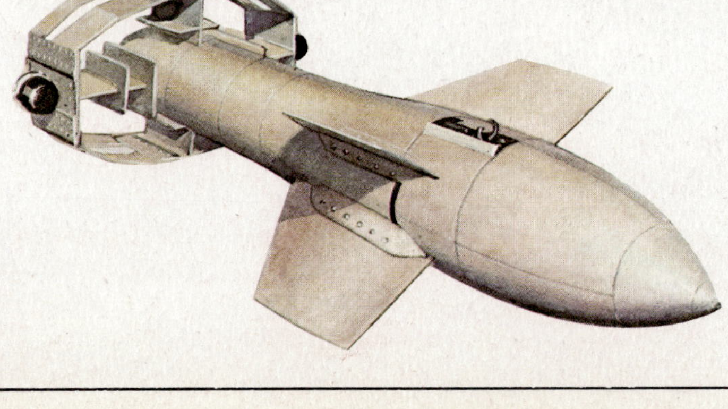

THE INFANTRY

At the beginning of World War II troops wore the same uniforms and had the same weapons as they did in 1914, and as far as tactics were concerned the Allies clung tenaciously to those they had been using 20 years previously. But the complicated manoeuvres and the constantly changing fronts brought about the use of automatic weapons.

Special groups of fighting units, known as guerrillas and commandos, were also brought into use, but these were only new in one sense. Because of their special role, the men in such groups used a new kind of weapon — the sub-machine gun, which was a personal, rapid-firing gun. This automatic weapon was extremely effective at 50m. The development of air attack also produced another kind of infantryman — the parachutist. It should be borne in mind that these special assault troops were used only when surprise was needed, and the traditional infantry carried out all the marching, occupying, and consolidating duties as usual.

Did you know . . .

. . . that *guerrilla* was originally a Spanish word, and that the first attacking operations of this kind took place during the Spanish War of Independence of 1808 against Napoleon's soldiers?

On the left are shown five uniforms used by soldiers fighting in World War II. All the colours were dull like those used in World War I, in order to camouflage the wearer and make it difficult for an attacker to pick him out against natural scenery and vegetation.

American Garand M-1 semi-automatic rifle

American semi-automatic carbine M-1

British Sten gun Mark III

Russian sub-machine gun PPSH, 1941

The United States was the only country that came into World War II fully equipped with semi-automatic weapons. The Russians and the Germans meanwhile were using their own traditional firearms. Sub-machine guns were extremely useful as assault weapons, and that is why they were used by every warring nation except Japan.

In addition to guns, the infantryman had another weapon of attack — the hand grenade (shown left). In order to increase the range of these weapons, the infantry had grenade-launchers and also miniature mortars. The grenades fired by these mortars were equipped with fins to stabilize their flight through the air (shown right).

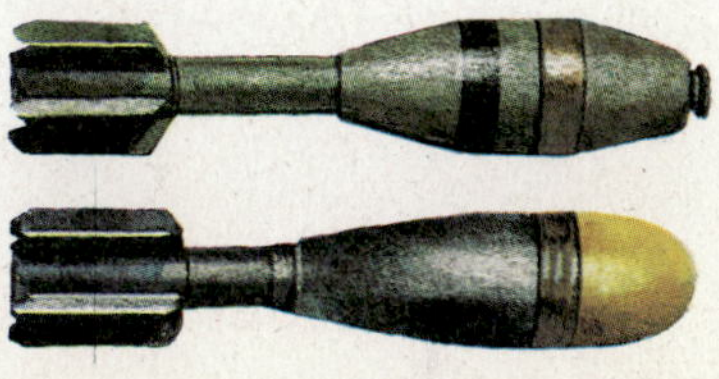

LANDMINES

'Mines!' There is a sudden explosion and a battle-hardened veteran lies dying. More than a destructive force, the mine is a very effective deterrent. Millions of mines were buried in the earth in World War II, and even today people are stumbling upon them.

There were three kinds of mines: anti-personnel mines, mines used against ordinary vehicles, and those used against tanks. They all exploded on contact, usually by being pressed upon. Each kind was a certain size and needed a precise weight to explode it. Sometimes these mines went off in unexpected ways. The mine could be fixed at the bottom of a hole and might blow up as it was being removed. Or it might detonate another mine some 10m behind it or by its side. There were all sorts of possibilities. Besides the straightforward underfoot mines, there were also booby-trap mines which went off when a door was opened, or when a bottle of wine was grasped, or when somebody lay down on a bed. A sign saying 'Danger, Minefield' (even if there was no minefield there at all) was enough to give pause to the bravest of men.

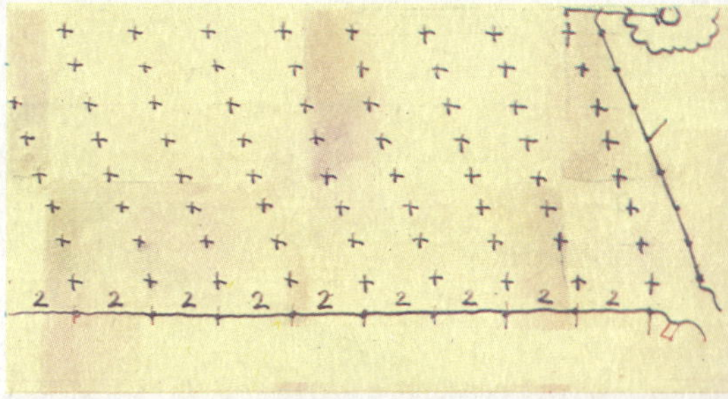

Minefields were laid with great care, making sure that each mine was a precise distance from its neighbour; and they were also mapped very carefully so that paths through minefields could be noted with great accuracy for the passage of friendly troops.

In order to detonate a minefield various mechanical contraptions were used. Sometimes a vehicle pushed two huge wheels weighing two tonnes before it. The wheels left tracks 90cm wide. Another device was a kind of flail that beat a path through a mine-field. The most efficient method was to send engineers and sappers into the minefield with magnetic detectors. When plastic or wooden mines were used, the ground had to be probed very gently with digging implements.

TODAY'S TANKS

It would appear that the modern tank, after having been seen in action in the Middle East wars between Israelis and Arabs, has reached its peak performance. In fact by the end of World War II, the tank was already dated because of its total dependence on air support. The German offensive in the Ardennes was successful so long as the sky was overcast; but as soon as the clouds cleared the tanks were smashed by Allied air attacks.

The pattern of today's tanks has already been set. It involves a weight of not more than 40 to 50 tonnes, a semi-automatic cannon of 120mm calibre, and powerful diesel engines that will carry the vehicle at a speed of 60-70km/h.

The future of tanks in warfare depends very much on the latest anti-tank measures, which are based on the use of rockets and precision artillery guided by laser beams.

NBT 70 tank

THE ATOMIC BOMB

It was only an experiment. Nobody knew exactly what was going to happen. On 16 July 1945 the first atomic bomb in the history of mankind was exploded — and it was a great success. The event took place in the desert wastes of Alamogordo, in New Mexico.

On 6 August of that year it was Hiroshima's turn; three days later Nagasaki's. Atomic bombs had been dropped on two Japanese cities. These two raids brought World War II to an end. The Germans had already surrendered, and now the Japanese followed suit. Humanity was entering, for good or ill, the Nuclear Age.

The Germans were the pioneers in nuclear fission in 1938, but the Americans were the first to produce an atomic bomb. A large group of scientists worked on the technical problems involved. They worked under the direction of Robert Oppenheimer (above left), a German-born professor of nuclear physics from the University of Berkeley, who has been called 'the father of the atomic bomb'. The actual testing of the bomb was carried out under the command of General Groves. Above is an artist's impression of the laboratory 'city' of Ark Ridge, Tennessee, where most of the research was done. On the right is a drawing of the tower on which the first atomic bomb was exploded in Alamogordo.

The raids on Hiroshima and Nagasaki produced countless victims. The latest estimate adds up to some 250,000 casualties. But more than the numbers of victims was the way in which they were stricken. Years after the event, many Japanese civilians were still dying horribly as a result of radiation, although few could have foreseen such terrible results at the time.

On the left is the 'Little Boy' atomic bomb dropped on Hiroshima. It was 3m long and weighed 4,100kg. Its nuclear warhead was made up of a few kilos of uranium-235.

The people who found themselves close to the point of impact were instantly atomized (no trace of them remained). Everything within a radius of 2km was completely burnt up. 3km and more away from the centre of the explosion, people received bad burns on their skin. After the explosion came the fire storm, caused by the enormous change in atmospheric pressure. And finally, there was the effect of radiation. At a distance of 3km from the centre of the explosion, many of the survivors were affected by neutrons and gamma rays. They lingered in agony for some time before dying.

The 'Fat Man' bomb dropped on Nagasaki. It had a plutonium warhead and weighed 4,500kg.

LIVING WITH THE NUCLEAR DETERRENT

There was no doubt that mankind had by now entered the Atomic Age, a period that was to be full of alarms and anxieties. The great fear was that men had still not learnt the lessons of the past and that they did not fully appreciate that the nuclear power at their fingertips could easily destroy their world.

When the secret of the atomic bomb became common knowledge, there came a frantic race to stockpile nuclear arms. This in turn led to the building of all kinds of expensive detection and early warning systems, with powerful radar screens and interceptor missiles for defence against sudden nuclear attack. The United States monitored the Russians, and vice versa.

This continuous state of tension, or cold war, on several occasions almost erupted into a hot war. But at the very last minute good sense prevailed in each case and the incident passed over harmlessly. Such an incident was the building of Soviet missile sites in Cuba, pointed at the United States, which almost started World War III.

THE HELICOPTER

The idea of vertical landing and take-off is as old as aviation itself. The genius of a Spanish inventor, Juan de la Cierva, who designed a practical rotor (large horizontal propeller), was what really made the advent of the helicopter possible. But helicopters did not become fully efficient machines until after World War II. In the Korean War the helicopter was a useful aid, but in the Vietnam War it became virtually the main weapon on the American side.

Above we see a Boeing Vartol in action. This took the place of lorries and jeeps for transporting troops because wheeled vehicles were unable to move in forested regions where there were no roads or tracks.

We might say, without being too far-fetched, that we live today in an era of aerial cavalry. Today's helicopters have proved themselves invaluable both in wartime and in civilian life.

TODAY'S WARPLANES

Quite apart from the invention of the jet engine, warplanes have developed in a spectacular way over a fairly short period. The latest types are truly impressive. The Lockheed SR-71, for example, flies at more than three times the speed of sound at a height of more than 24km. The Galaxy, a military transport, can carry a load of 100 tonnes at a cruising speed of 815km/h.

We cannot tell what the future holds. What seems to be quite clear is that the days of military aircraft are numbered. Their fate may resemble that of the artillery: although big guns lost their dominant role they did not for that reason disappear from the battlefield. Actually, aircraft have for many years already been replaced by another kind of airborne artillery — the missile.

Below is an enormous B-52 Stratofortress (fortresss of the stratosphere), a 200-tonne bomber used by the Americans in the Vietnam War on conventional bombing raids. Because of new kinds of warfare now being contemplated, this is the last aircraft of its kind in the history of military aviation.

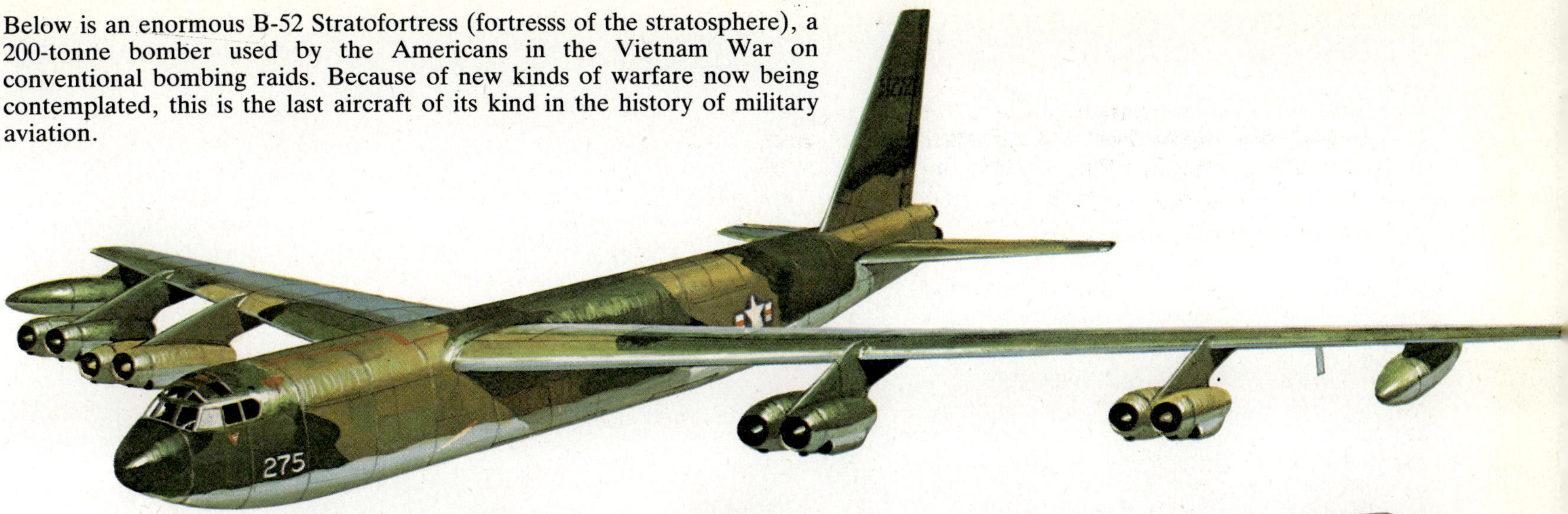

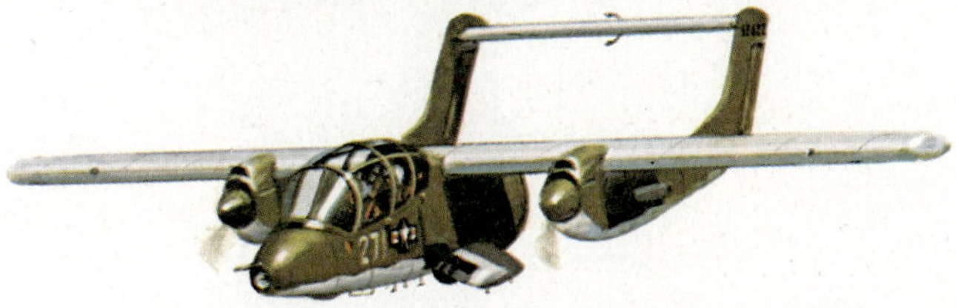

In spite of their liking for speed, the Americans in Vietnam pressed into service ancient propeller-driven aircraft that were used successfully at low altitudes. The picture above shows a Bronco OV-10, which was used for attacking sorties.

Above is the SR-71, one of the fastest warplanes in service to date. It was designed in 1961 to replace the American U2 spy planes. It is able to operate at great heights and is equipped with missiles.

Because modern aircraft became so costly to make, designers tried to make single types perform as many jobs as possible. One way of doing this was to design an aircraft that could change the shape and angle of its wing in flight and thus adapt itself to a variety of speeds without loss of power, manoeuvrability, or altitude. A problem arose as to where to locate the guns, because most fighters carry them under the wings. Above and right is the French Mirage fighter showing the wings spread at two different angles.

Below is a picture of the first and only fighter in regular service that can take off and land vertically. It can fly at the speed of sound and land like a helicopter. It is the British Hawker Harrier and the secret of its success lies in its Rolls Royce Pegasus turbo jet.

ROCKETS AND ELECTRONICS

Everybody agrees that the rocket is the weapon of today and of the future. All the armies of land, sea and air are now equipped with rockets, and these are gradually replacing most traditional weapons.

The great advantage of a rocket-powered missile is that it can be guided, and its path can often be corrected in flight in order to make sure that it reaches the target. This has been made possible by means of electronic 'brains'. Computers, radar chains and spy satellites all contribute to the wonderful accuracy of long-range missiles, whether they are carrying one or more warheads.

The missile has already proved its worth in conventional warfare. The most popular kinds are those that are remotely or immediately controlled by radar. Some are equipped with TV cameras; others have infra-red detectors or laser beams to help them home in on their targets. The defence against missile attack consists mainly in destroying them in the air after having detected them by radar and intercepted them by means of calculations worked out by computer. In addition, a number of 'scrambler' devices have been developed to confuse the missile guidance systems and their detector gadgets.

Intercontinental missiles are stored away in vast underground chambers, which are more or less small, self-complete townships. These underground defensive systems form a chain which, apart from the distances involved, reminds one of the great fortified lines of World War II.

The picture above shows a triple launcher of Hawk ground-to-air missiles. The missiles are guided by radar, and their solid-fuel propellant gives them great velocity.

Above is a three-tonne missile known as Honest John. It is a ground-to-air missile, equipped with a nuclear warhead, and replaces conventional artillery.

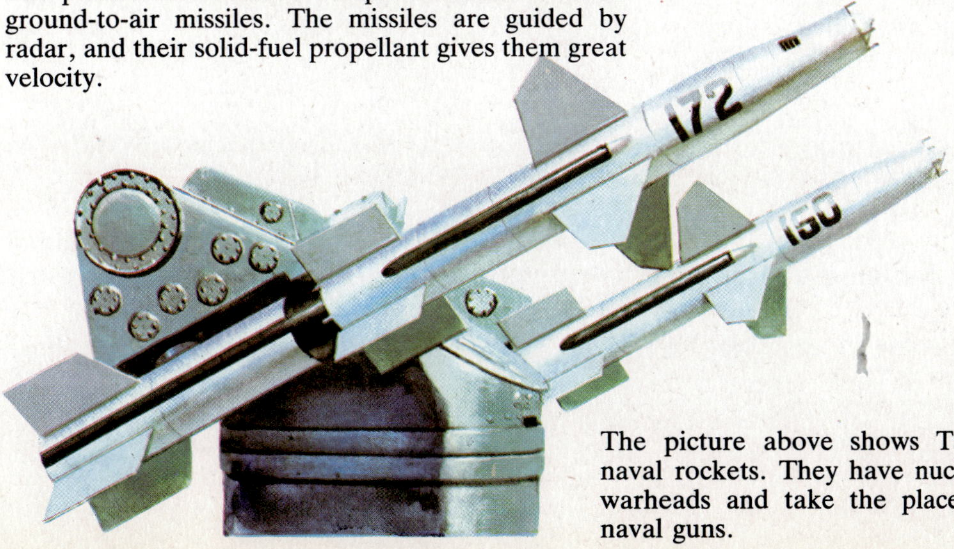

The picture above shows Talos naval rockets. They have nuclear warheads and take the place of naval guns.

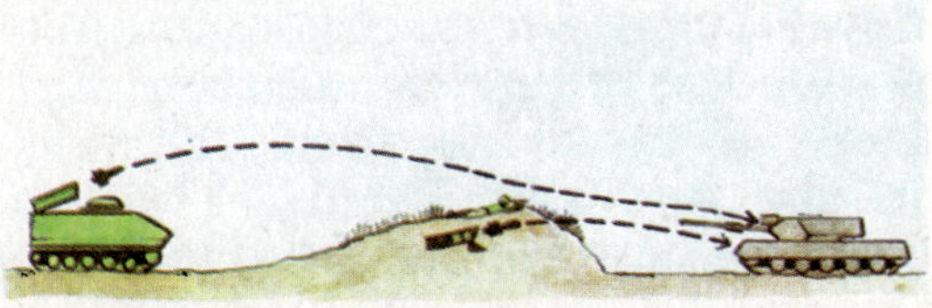

Above we see how an anti-tank missile achieves its target.

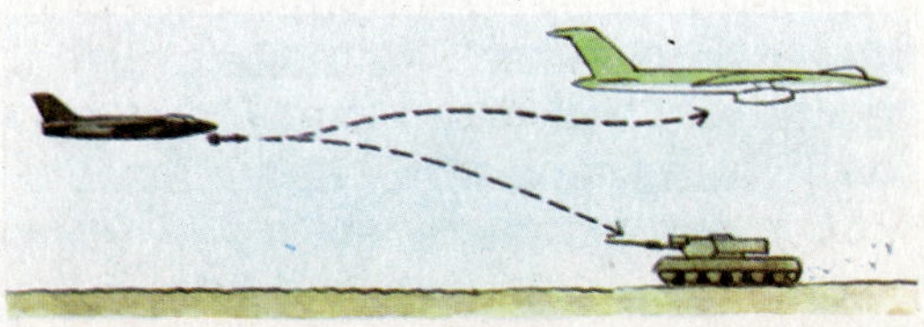

The drawing above shows how air-to-air and air-to-ground missiles work.

WHAT TOMORROW HOLDS

Nobody knows what will happen to mankind in the distant future, but we can speak with some confidence about the years immediately ahead because new weapons for use then are being developed and tested under our noses. The main war efforts will not be to seek new and better explosives because it seems that mankind has already found the most destructive weapon possible in the hydrogen bomb. The main problems worrying the planners now are improved means of defence, and how to make the best of an attack using nuclear bombs with a destructive capacity of between 1,000 and 50,000 tonnes of TNT. In addition to nuclear missiles, wars will be waged with artificial satellites and laser rays. Huge air fleets will be equipped with missiles, anti-torpedo craft, miniature aircraft carriers, etc. The use of computers will mean that fewer people will actually take part in the fighting, and army commanders will be able to enjoy a front seat at the action by means of television.

The only conventional force that will still be employed in battle, as it has been for thousands of years, will be the infantry. But this will be equipped with new kinds of weapons to suit the occasion.

CONCLUSION

It seems strange and ridiculous that up to now man has still not been able to avoid being involved in wars of one kind or another. It is even odder to realize that such wars are generally entered into after a great deal of thought, and not declared on the spur of the moment. But often the result of the fighting has not been given enough attention. The truth is, that almost all wars between nations have always been started by a single person, or perhaps a small group of people, with a particular axe to grind. The great majority of people making up the warring nations have no interest in such conflicts, but are eventually dragged unwillingly into the terrible violence that regrettably continues to erupt in various parts of the world to this day.

Nevertheless there exist among the ordinary masses of mankind, a great many good, moderate people who today are linked more closely than ever by means of modern communications. These are people of goodwill whose peace-loving ideas cut across national boundaries and politics to bring real hope for the survival of humanity.

17952
U.S. AIR FORCE
17A
LT231
G209